AMERICAN INSTITUTE OF CHEMICAL ENGINEERS
EQUIPMENT TESTING PROCEDURES COMMITTEE

Chairman: Thomas H. Yohe
Swenson Process Equipment

Vice Chairman: Dr. P.C. Gopalratnam
E.I. DuPont de Nemours & Co.

CENTRIFUGAL PUMPS (Newtonian Liquids)
PROCEDURES SUBCOMMITTEE

Chairman: Robert J. Hart
E.I. DuPont de Nemours & Co. (Retired)

Subcommittee Members

Gary Clasby
Flowserve Corporation
Dayton, OH

Allen Budris
Goulds Pumps – ITT Industries
Seneca Falls, NY

Roy Dobson
Fluor Daniel
Sugarland, TX

Van Wilkinson
Shell Oil
Houston, TX

AMERICAN INSTITUTE OF CHEMICAL ENGINEERS
EQUIPMENT TESTING PROCEDURES COMMITTEE

GENERAL COMMITTEE
(2001)

Thomas H. Yohe, *Chairman*
Swenson Process Equipment

Dr. P.C. Gopalratnam, *Vice Chairman*
Anthony L. Pezone, *Past Chairman*
Dr. Soni O. Oyekan, *Council Liaison*
Mark Rosenzweig, *Staff Liaison*

GENERAL COMMITTEE MEMBERS

Prashant Agrawal
S. Dennis Fegan
Dr. John G. Kunesh
J.H. Hasbrouck
Gary O. Oakeson
Richard P. O'Connor
Robert E. McHarg

CONTENTS

100.0 Purpose and Scope

101.0 *Purpose*

101.1 This testing procedure provides methods of conducting and interpreting field tests on centrifugal pumps with actual pumped fluids. Measuring methods practical for field test conditions are presented to guide the user. These tests may be desired for the following purposes:

101.1.1 Checking a new pump for operability, to determine if a pump will function both mechanically and hydraulically to do the job for which it was intended.

101.1.2 Testing a pump for overall system performance, to determine if the pump will deliver all the fluid that it was designed to deliver under various load conditions as demanded by the process.

101.1.3 Testing to establish operating characteristics, such as total head, capacity, and power requirements. Net Positive Suction Head Required may be determined as part of this procedure when test code accuracy (see section 102.2) is not required. Such measurements can also help resolve operating problems.

101.1.4 Information obtained when testing a pump as part of a regularly scheduled maintenance program may be used to evaluate the mechanical condition of the pump.

101.1.5 Periodic testing to establish a performance history which can be used to monitor a change in the pumping conditions, and energy consumption. Performance testing can be used to determine if a change in the process has occurred.

101.2 Refer to 805.1 for procedures used by many manufacturers to test new equipment at the point of manufacture from which performance curves are published.

102.0 *Scope*

102.1 This procedure recommends methods for conducting and reporting field tests on centrifugal pumps handling Newtonian fluids.

102.2 It should be noted that the preferred method of establishing pump performances is under controlled test facility conditions. Testing can generally be accomplished by the pump manufacturer at the factory and may be witnessed by the purchaser. It should also be noted that this field test procedure should not be used when test code accuracy is required. For details on test code accuracy (see reference 805.1) or American Society of Mechanical Engineers Centrifugal Pumps performance test code, (see reference 805.3) that defines accuracy required of such tests.

102.3 Safety and environmental concerns are listed in Sections 301.1 and 301.2.

103.0 *Liability*

AIChE and members of the various committees involved make no representation, warranties or guarantees, expressed or implied, as to the application or fitness of the testing procedures suggested herein for any specific purpose or use. Company affiliations are shown for information only and do not imply procedure approval by the companies listed. The user ultimately must make his own judgment on the specific testing procedures he wishes to utilize for specific applications.

200.0 Definitions and Descriptions of Terms

201.0 *Definition*

This section contains the definition of all terms used in this testing procedure. Nomenclature and typical units are also listed in Section 801.0.

202.0 *Centrifugal Pumps*

A pump in which pressure is developed principally by the action of centrifugal force.

203.0 *Newtonian Fluids*

A liquid which resists shear in direct proportion to the rate of shear.

204.0 *Datum*

The reference from which all elevations are measured.

205.0 *Specific Gravity, (s)*

The dimensionless ratio of the specific weight of the liquid to the specific weight of water under reference conditions of temperature and pressure. It is primarily a function of temperature, but operating pressure and elevation will have some effect on specific gravity and should be evaluated.

206.0 *Vapor Pressure*, (h_{vpa})

The absolute pressure exerted by the vapor of a liquid at a given temperature.

207.0 *Viscosity*, (μ)

The measure of a fluid's tendency to resist an internal shearing force.

208.0 *Pressure Head*, (h_p)

The pressure energy per unit weight of liquid at a given location corrected to datum elevation.

209.0 *Total Discharge Head*, (h_d)

The reading of a pressure gauge at the discharge of the pump, converted to column of liquid and referred to datum, plus the velocity head at the point of gauge attachment.

210.0 *Total Suction Head*, (h_s)

The reading of a pressure gauge at the suction of the pump, converted to column of liquid and referred to datum, plus the velocity head at the point of gauge attachment.

210.1 *Total Suction Lift, (h_{gs}):* Suction lift exists where the total suction head is below atmospheric pressure. Total suction lift, as, determined during testing, is the reading of a liquid manometer at the suction nozzle of the pump, converted to column of liquid and referred to datum, minus the velocity head at the point of gauge attachment.

211.0 *Total Head*, (H)

The measure of the work, ft lb./ lb. of the liquid, imparted to the liquid by the pump. Therefore, pump total head is the algebraic difference between the total discharge head and total suction head.

212.0 *Static Discharge Head*, (h_{sd})

The portion of the discharge pressure exerted on the pump, referred to datum and converted to column of liquid pumped, that is due to the net liquid elevation above the pump discharge plus the pressure that is acting on this liquid surface is defined as the Static Discharge Head. This pressure can be measured with a discharge pressure gauge and without the pump running.

213.0 *Static Suction Head*, (h_{ss})

The hydraulic pressure exerted by the liquid against the pump suction as indicated by a pressure gauge, manometer or similar devices, referred to datum and converted to column of the liquid being pumped, without the pump running.

214.0 *Suction Surface Pressure*, (h_{ps})

The pressure exerted on the surface of the liquid at the suction sources of the pump.

215.0 *Discharge Surface Pressure*, (h_{pd})

The pressure exerted on the surface of the liquid in the vessel to which the pump discharges.

216.0 *Friction Head Loss*, (h_f)

The resistance to flow in a pumping system as indicated by loss of head. Suction Friction Head will be designated as h_{fs}. Discharge Friction Head will be designated as h_{fd}.

217.0 *Gravitational Acceleration*, (g)

The gravitational constant for customary unit is 32.2 ft/ sec^2 and 9.81 m/ sec^2 for SI units at sea level.

218.0 *Velocity Head*, (h_v)

The energy component in a pumping system that represents the kinetic or "velocity" energy in a moving liquid at the point being considered in the system. It is equivalent to the vertical distance the mass of liquid would have to fall (in a perfect vacuum) to acquire the velocity "v" and is expressed as:

$$h_v = \frac{v^2}{2g}$$

Suction Velocity Head will be designated h_{vs} and Discharge Velocity Head will be designated h_{vd}.

219.0 *Pump Efficiency*, (eff)

The output (hydraulic) horsepower supplied by the pump divided by the input (shaft) horsepower expressed as a percentage.

220.0 Net Positive Suction Head Required, ($NPSH_R$)

The $NPSH_R$ is defined as the NPSH applied to the pump at a given flow rate that causes sufficient cavitation to reduce the Total Head by 3%. This is an established pump industry standard procedure (see reference 805.1) used to indirectly measure, at a reasonable cost, the suction side pressure loss inside a pump before mechanical action increases the liquid pressure. The measurement is taken while pumping water with a minimum of inlet stream turbulence (i.e. no close connected double elbows), no entrained gas, and frequently with water that has most dissolved gas removed.

220.1 Discussion and considerations for $NPSH_R$

220.1.1 *Net Positive Suction Head Required* ($NPSH_R$) is a crucial performance limitation of any centrifugal pump. It must be considered with equal importance to head, capacity, and discharge pressure when designing a pumping system and selecting a pump for a given service. The $NPSH_R$ of a given pump is a function of the impeller design, diameter and speed, and is established by specific and well-defined procedures by test using water as the testing fluid. A pump, in order to operate reliably and with the minimum of maintenance, must be supplied with a $NPSH_A$ (defined below) in excess of $NPSH_R$ value. Reference 805.1 describes more detailed information on this subject and the appropriate margin between $NPSH_A$ and $NPSH_R$. Performance test results will be meaningless if adequate margin does not exist.

220.1.2 *Net Positive Suction Head Available,* ($NPSH_A$): A detailed calculation of $NPSH_A$ should have been done prior to purchasing the pump. Preparations for field-testing should determine adequate $NPSH_A$ to $NPSH_R$ margin actually exists prior to conducting a test. Insufficient margin will reduce pump hydraulic performance compared to published test curves supplied by the manufacturer. Excessive turbulence, entrained gas and liquid with large quantities of dissolved gas will decrease the $NPSH_A$ values obtained during the test.

220.1.3 $NPSH_A$ equals the suction surface pressure at the liquid surface, minus the vapor pressure, plus the suction static head (all in column of liquid) minus the friction loss through the suction system. If suction pressure is measured at the pump suction flange, the friction loss (h_{fs}) value is zero.

$$NPSH_A = h_{psa} + h_{ss} - h_{fs} - h_{vpa}$$

h_{psa} = suction surface pressure, absolute
h_{ss} = static suction head (plus or minus depending on actual conditions)
h_{fs} = suction friction head
h_{vpa} = vapor pressure of liquid, absolute, at temperature.

220.2 Two common $NPSH_A$ situations are shown in Figures 220.2.1 and 220.2.2.

220.2.1   Pressurized tank on suction (see Figure 220.2.1)

$$NPSH_A = const \frac{h_{psa} - h_{vpa}}{s} + h_{ss} - h_{fs}$$

Where:
const = 2.31 for customary units to convert PSI to feet of liquid
const = 0.102 for SI units to convert kilopascals to meters of liquid

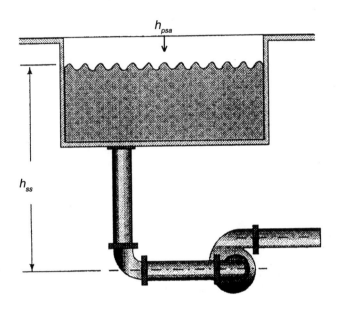

Figure 220.2.1 Pressurized tank on suction. $NPSH_A$ is measured at pump center line and face of inlet flange.

220.2.2   Suction lift (see Figure 220.2.2)

$$NPSH_A = const \frac{h_{psa} - h_{vpa}}{s} - h_{ss} - h_{fs}$$

Where const = 2.31 for customary units
        const = 0.102 for SI units

Note: h_p equals the atmospheric pressure on an open tank.

220.2.1   Pressurized tank on suction (see Figure 220.2.1)

$$NPSH_A = const\ \frac{h_{psa} - h_{vpa}}{s} + h_{ss} - h_{fs}$$

Where:
const = 2.31 for customary units to convert PSI to feet of liquid
const = 0.102 for SI units to convert kilopascals to meters of liquid

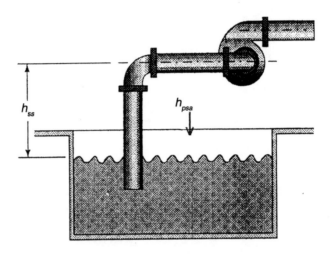

Figure 220.2.1 Pressurized tank on suction. $NPSH_A$ is measured at pump center line and face of inlet flange.

220.2.2   Suction lift (see Figure 220.2.2)

$$NPSH_A = const\ \frac{h_{psa} - h_{vpa}}{s} - h_{ss} - h_{fs}$$

Where const = 2.31 for customary units
const = 0.102 for SI units

Note: h_p equals the atmospheric pressure on an open tank.

220.3.1 Manufacturers expect the user to supply an $NPSH_A$ that exceeds the published $NPSH_R$ value. The margin between the $NPSH_A$ and $NPSH_R$ becomes a commercial decision and should be properly evaluated by the person selecting the pump and developing the piping system. See paragraph 220.1.1 & reference 805.1

220.3.2 Anyone developing pumping systems designed with these minimum recommended margins should consider testing the equipment to be delivered to confirm that it meets the published data (see reference 805.1). The following information should be considered when ordering the pumps to decide if tests should be conducted:

220.3.2.1 Most manufacturers do not hydraulically test every pump unless required to do so by the purchaser. General pump curves reflect typical performance for a series of pumps. For critical applications it may be prudent for the purchaser to witness the test.

220.3.2.2 The "first" test of a specific pump may result in higher $NPSH_R$ and Total Head values than indicated by the published data due to casting variations of the casing and impeller. To correct this condition, the manufacturer may have to grind the controlling surfaces of the impeller and case to reduce the $NPSH_R$ and reduce the impeller diameter to reduce the total discharge head, h_d within acceptable tolerances. After such modifications are made, the pump is retested to confirm the results of the rework. Test specifications should be included in the purchase order when it is deemed necessary to test a pump.

220.3.2.3 Published $NPSH_R$ values are not guaranteed by the supplier unless so specified by the purchaser at the time the order is placed and the pump is $NPSH_R$ tested.

220.3.2.4 The testing procedure followed by most manufacturers (see reference 805.1) typically yield the minimum information, which is not considered adequate for most critical chemical plant services, especially as it relates to $NPSH_R$ testing. Centrifugal pump test specifications may be made part of purchase specifications.

220.3.2.5 Cavitation that affects the total head developed by the pump starts at values well above the published $NPSH_R$ values.

220.3.2.6 $NPSH_A$ values only slightly less than the $NPSH_R$ values will cause a dramatic loss in pump head, which causes the flow in an actual system to surge and the systems to become hydraulically unstable. During this type of operation, pump vibration increases causing increased equipment maintenance. Actual fluid and piping conditions will normally decrease the $NPSH_A$.

220.3.2.7 The $NPSH_R$ established in these tests will never be less than that in actual operation. Also restricting clearances in the pump wear (impeller and casing wear rings), will increase the $NPSH_R$.

300.0 Test Planning

301.0 *Conditions*

301.1 *Safety:* Any equipment testing must conform to the latest requirements of all applicable safety standards. These include but are not limited to plant, industry, local, state, and federal regulations. It is recommended that all testing be conducted under the supervision of personnel fully experienced in plant and equipment operating practices.

301.2 *Environmental:* The test procedures must conform to the latest requirements of all applicable environmental standards, which include plant, industry, local, state, and federal regulations. Environmental conditions that apply to the equipment in normal operation should also apply during testing.

302.0 *Resources*

302.1 *Time:* The test should be scheduled at a time when suitable operating conditions can be obtained and the presence of key personnel to carry out the test can be assured. Time should be allowed to determine objectives and provide proper instrumentation before an actual test is performed.

302.2 *Equipment:* After determining which measurements are necessary for achieving the test objectives, equipment or instrumentation must be selected and installed to obtain the necessary measurements. The required precision or accuracy of the measurements is an important factor in determining which instruments should be used. Also, proper consideration should be given to calibrating the instruments and installing them according to recognized procedures or practices.

303.0 *Process Considerations for Performing Tests*

If the test is to be performed under actual process conditions, as in an operating production unit, attention should be given to possible process upsets. A test may require the flow be "dead headed" (no flow) or that maximum flow rate is obtained.

303.1 For pumps that are being tested due to major flow discrepancies, the following steps may be considered for evaluating the system, when a means of measuring flow is not available.

303.2 If the pump and the system will permit a momentary deadheading (no flow while the pump is running) the head developed by the pump can be established providing a point on the test performance curve at zero flow.

303.3 If the resulting head compares to the manufacturers published curve, within +/- 10%, it is reasonable to assume the pump will provide the published values for the remaining flows and other system related issues are causing the discrepancy.

303.4 Also, suction and discharge pressures may be varied during a test. Determination must be made if the process can tolerate large changes in conditions, such as flow and pressure, without having the process gets out of control. Operating the pump at various flows can also change the temperature of the liquid being pumped. Having to operate the pump at conditions other than at design can cause mechanical problems that could result in a sudden shutdown of the pump, i.e., motor overload. Knowing the capability of the pump and the process system is essential in setting up the test to avoid unwanted upsets.

304.0 *Data Requirements*

The data required to evaluate the performance of a centrifugal pump depend on the purpose of the test. The following data may be required.

304.1 *Field Test Data*

304.1.1 Suction pressure at or near the suction flange of the pump, corrected for the gauge elevation above or below the selected system datum plane, which is usually the centerline of the pump.

304.1.2 Discharge pressure at or near the discharge flange.

304.1.3 Power Consumption: Electric motor Amps, electric motor kilowatt usage, steam rate of turbine, or fuel usage of engine.

304.1.4 Capacity (volume rate of flow) at the pump.

304.1.5 Speed of the pump shaft in revolutions per minute.

304.1.6 Liquid temperature at the suction and discharge.

304.1.7 Barometric pressure.

304.2 *Installation Data:* The following data concerning the installation should be known:

304.2.1 Piping details: items such as valves, orifices, reducers and suction strainers will affect performance.

304.2.2 Proper location of instrumentation.
Difference in elevation between the gauges and the centerline of the impeller must be considered.

304.3 *System Data or Equipment Information*

304.3.1 The required pump Brake Horsepower is an integral part of pump testing and must be evaluated. As electric motors are the major drivers for pumps see reference 805.2 for more details in estimating motor output horsepower to the pump. It is beyond the scope of this procedure to address all forms of drivers.

304.3.2 Pressure gauge calibration.

304.3.3 Suction and discharge pipe inside diameters.

304.3.4 Liquid Characteristics: specific gravity, viscosity and temperature effects. Specific gravity, vapor pressure and viscosity measurements are not within the scope of this procedure. These characteristics should be obtained from analyses of process samples or technical reference books.

304.3.5 Pump Information: rated capacity, rated head, rated speed, impeller drawing number, impeller diameter, materials of construction, mechanical seal or packing details, $NPSH_R$, efficiency, and performance curve.

305.0 *Pump Considerations*

Certain conditions should be observed and recorded prior to the initiation of a pump performance test.

305.1 The pump should be identified, preferably by serial number. All available information concerning the pump's performance should be obtained. This information will be used later in the test.

305.2 The pump history should be known. A newly installed pump would be tested in a different manner than an older pump. If the pump has had problems, it should be noted prior to the test so that considerations can be given to the effect of the problems on the test results.

305.3 *Pump Installation*

305.3.1 Alignment between pump and driver should be checked.

305.3.2 The installation should be inspected for proper support of the pump. If the support is not adequate, it can cause misalignment between the pump and driver. Misalignment can cause vibration, bearing and seal failure, and coupling wear.

305.3.3 The foundation conditions should be noted at this time. Does the pump appear to be on a substantial concrete foundation? Is it mounted on a substantial frame? Are there externally excited vibrations in the foundation area from the nearby equipment? These factors should be noted at this time for later evaluation.

305.3.4 The pump piping system should be observed. Conditions such as long lengths of unsupported pipe, thermal expansion, etc. may impose excessive strain on suction and/or discharge nozzles of the pump. Excessive strain could cause misalignment, internal rubbing and affect the performance of the pump.

305.3.5 If a suction strainer is used, the cleanliness and the location will affect pump performance.

305.4 *Hydraulic Conditions*

305.4.1 Operational conditions of the pump should be considered. It is to be noted that there may be flow limitations. Centrifugal pumps may have minimum flow limitations due to heat rise, internal recirculation or mechanical limitations. There may also be maximum flow operational limits. Some pumps are designed to run with an open discharge valve while others may require discharge valve throttling to keep them at an acceptable point on their performance curve. These possibilities should be checked with the manufacturer.

305.4.2 The suction conditions of the pump should be observed. The suction piping system should be considered to insure proper inlet flow conditions. Observations of suction vessel level should be conducted during the pump test to assure smooth inlet flow. Observation of entrained gas or vortexing should be noted.

305.4.3 In taking performance data, there may be fluctuations in the readings due to pump instability, system instability, or mechanically generated vibrations. These should be identified and corrected before proceeding with the test.

305.4.4 It may be necessary to take data readings simultaneously, such as inlet and outlet temperatures. This may require special provision if instrumentation is separated from the pump.

305.5 *Conditions That May Affect Performance*

305.5.1 Gas in the fluid will affect the performance of a centrifugal pump. The presence of 4% entrained vapor in a fluid can cause a reduction in pump total head at a given capacity in excess of 40%. Even small amounts of gas may substantially increase the $NPSH_R$ of the pump reducing or even eliminating the anticipated margin between the $NPSH_A$ and $NPSH_R$.

305.5.2 Submergence of the suction pipe on a horizontal or vertical pump is critical to proper performance. Insufficient submergence can induce vortexing or other problems that can affect pump performance.

305.5.3 The pump and the system involved in the performance test should be checked for leakage. Excessive leakage can cause erroneous performance indications. The piping should also be checked to determine if there are any bypass lines that may divert flow from the flow measuring device. It should be noted that a line with an apparently closed valve could still be diverting fluid from the flow-measuring device if there is internal damage to the valve.

305.5.4 The pump stuffing box should be checked. Stuffing box packing that has been tightened excessively can cause erroneous power reading on smaller pumps. The same condition can exist with improperly adjusted mechanical seals. Turning the pump over by hand prior to operation can give an indication of such a problem.

305.5.5 Manufacturer's published test curves typically do not include the power losses of mechanical seals or shaft packing as equipment is normally tested at a low suction pressure with only a minimal shaft seal resistance. High suction pressures and double mechanical seals can add a measurable load to pump driver of small pumps (with drivers of 25 hp or less).

305.6 *Internal Mechanical Condition:* The wear surface clearances, impeller adjustments, and general internal wear can affect performance. The troubleshooting section (see section 703.0) should be used to diagnose problems.

306.0 *Test Plan*

A test plan should be prepared taking into account the foregoing considerations and the conditions of the specific system and equipment.

400.0 Instrumentation and Measurement Methods

401.0 *Introduction*

This section presents detailed information on instruments and methods of measurement most commonly used in the field testing of centrifugal pumps.

401.1 For pump testing, it is necessary to determine:

401.1.1 Pressure (suction, discharge, barometric), corrected for difference in elevation of pressure gauge and selected datum plane.

401.1.2 Temperature (liquid, and local ambient air),

401.1.3 Rate of flow,

401.1.4 Pump speed,

401.1.5 Driver power input.

401.2 For the liquid being pumped:

401.2.1 Specific gravity,

401.2.2 Viscosity,

401.2.3 Vapor pressure.

402.0 *Accuracy and Precision*

Field test results are dependent on the accuracy of the measurements and assumptions. Care must be given in measuring those items that would affect the accuracy of the testing variables before challenging actual test stand results.

402.1 *Accuracy*

Instruments or measuring devices should be calibrated over the expected operating range against known standards. The accuracy of all data should be determined, and the effects on the test results considered.

402.2 *Precision*

Instruments should be selected that will measure the variable at the point of highest precision, if at all possible. This is frequently the mid-range of the selected instrument. The precision of an instrument is usually reflected by its smallest scale graduations. Instrument readings recorded should reflect this precision.

403.0 *Measurement of Test Parameters*

403.1 *Instruments and Indicating Devices* suitable for this test are given in the following.

403.2 *Pressure Measurements*

403.2.1 The measurement of pressure for the determination of head is carried out by pressure indicating devices (columns, manometers, gauges or transducers) connected with the liquid passage through pressure taps. In some cases, the measurement is obtained through pressure transmitters which are in turn connected to the liquid passage through pressure taps. When transmitters are used, the transmitter and the pressure-indicating device are considered as one complete instrument and are to be calibrated as a unit.

403.2.2 All gauge line connections must be tight. All instrument hose, piping, fittings, valves, etc. should be checked under pressure prior to the test to assure that there are no leaks. Connecting lines should be vented to remove all gas.

403.2.3 The suction head could be approximated by measuring the distance the level of the suction liquid is above or below the centerline of the pump suction minus the calculated losses in the suction piping plus the suction surface pressure.

403.2.4 Pressure Taps

403.2.4.1 Pressure taps in the pipe should be flush with and normal to the wall of a cylindrical portion of the liquid passage. The interior edge of the pressure tap hole should be free of burrs and roughness, as these will cause turbulence, which may interfere with obtaining a true pressure reading. Pressure taps in or within five pipe diameters downstream of an elbow should be avoided.

403.2.4.2 Pressure readings may vary at different points on the periphery of a pipe, owing to disturbances in the flow pattern caused by internal roughness, a close proximity to fittings, a misplaced gasket, etc. Therefore, the pressure should preferably be determined at a location where flow disturbances are expected to be minimal.

403.2.4.3 Many materials can cause difficulties or inaccurate readings by corroding, plugging and freezing or condensing in these pressure connections at atmospheric temperatures or operating pressures. To prevent errors it may be necessary to use a buffer fluid between the measuring device and the operating fluid. If the specific gravity of the buffer fluid is different than the that of the fluid being pumped, the measurements will have to be adjusted to reflect this difference. Measuring devices may also be isolated by diaphragms.

403.3 *Temperature Measurement*
If temperature variations during the test have substantial effect on fluid viscosity or density, special consideration will be applied to accommodate these changes.

403.3.1 Temperature may be measured by using one of the measuring devices listed below:

403.3.1.1 Liquid-in-glass thermometer,

403.3.1.2 Bimetallic dial thermometer,

403.3.1.3 Thermocouples,

403.3.1.4 Resistance Temperature Device (R.T.D.).

403.3.2 The liquid-in-glass and bimetallic dial thermometers are the least expensive and are direct reading. The thermocouple and resistance temperature devices require a potentiometric instrument for readout. The selection of which measuring device to use depends upon accuracy, cost, installation, and safety considerations.

403.3.3 Generally temperature variations are not crucial to test results. If temperature variations affect fluid characteristics, special considerations may be required.

403.4 *Capacity Measurement*

403.4.1 The measurement of capacity may be carried out by rate-of-flow meters, weighing, or volume methods. The reliability as well as the accuracy of weight or volume methods makes them preferable to rate-of-flow meters because the calibration of many rate-of-flow meters varies with viscosity.

403.4.2 By Weight: The measurement of capacity by weight depends upon the accuracy of the scale used and the measurement time. This method involves determining the change in the gross weight of a vessel on the suction or discharge of the pump in a given period of time.

403.4.3 By Volume: This method involves determining the volume of liquid being transferred to or from a vessel of known dimensions over a given period of time. Volume may be determined by measurement of the change in liquid level in a vessel. In this case, the liquid should be introduced into the tank with a diverter, and suitable baffles should be used in the tank to reduce swirls and eliminate surface effects.

403.4.4 By Venturi Meters, Orifice Meters, and Nozzles: Venturi meters, square-edged concentric orifice plates, and circular nozzles of the convergent type (both the submerged-flow and the free-discharge type) are acceptable instruments for field testing of centrifugal pumps. The installation and sizing of these meters should be in accordance with the manufacturer's recommendations. Selection of these meters should be based on accuracy, cost, installation, and safety considerations.

403.4.5 By Other Instruments: Rotameters, displacement meters, Doppler-effect meters and magnetic flow meters, and Pitot tubes may be used. However, they should be calibrated by a weight or volume technique using the test liquid at conditions of rate of flow and temperature to be encountered in the test. The instrument manufacturer should be consulted for test liquid effects.

403.5 *Speed Measurement:* The pump speed should be measured by any reliable revolution counter or a speed-measuring device, such as stroboscope, handheld tachometer, electronic counter with pulse generator or vibrating-reed tachometer.

403.6 *Power Measurement*

403.6.1 The majority of centrifugal pumps in the field are driven by a/c electrical motors. The power input to a direct-connected pump can be estimated within the accuracy of this procedure by using the electrical power in watts and multiplying by the operating point efficiency of the motor to determine the power input to the pump shaft.

403.6.2 When the overall efficiency of a pump and driver is to be determined, this procedure shall apply to the determination of the pump performance only. The appropriate test procedure for the driver should be referred to in determining driver input.

403.6.3 The power input to the motor can be measured at the motor terminals by any one of the following acceptable methods:

403.6.3.1 Polyphase wattmeter,

403.6.3.2 Single-phase wattmeter,

403.6.3.3 Voltmeter, ammeter and power factor where required.

403.6.4 Power input from other drivers, such as steam turbines or hydraulic motors, must be obtained in cooperation with the manufacturers of such equipment.

500.0 Test Procedure

501.0 *Pre-Run Data*

The pre-run data as decided by following Test Planning Section 300.0 should be obtained before the test runs are begun.

501.1 Obtain, or determine and record the properties of the liquid being pumped during the test.

501.2 Measure and record the measurements decided upon by the plan; for example, suction entry to discharge centerline on sump pumps.

501.3 Record the driver data.

501.4 Calibrate all instruments involved in the test as outlined in the appropriate portion of Section 400.0.

502.0 *Pretest Checks*

Prior to operating the pump to perform the test the following checks should be made.

502.1 Check that the physical requirements as developed under Planning have been completed - for example, alignment.

502.2 Determine that the fluid to be pumped is available in sufficient quantities to complete all the test runs to be made.

502.3 Follow and check the piping systems to insure that fluid will be delivered to, and be discharged from the pump as required by the needs of the test. Are there branch lines to be accounted for in obtaining flow, etc.?

502.4 Check the driver and pump for direction of rotation, freedom of turning and coupling alignment

502.5 Check the instrumentation to be sure sensing systems are full and are not influenced by ambient conditions.

503.0 *Trial Run Checks*

503.1 Open the pump suction and discharge valves. In some cases it may be desirable to begin with a partially open discharge valve to avoid the effect of full flow surge on some system components or to reduce power requirements at start-up. Make sure the pump is primed.

503.2 Start the pump by applying power or admitting steam to the driver.

503.3 Observe this initial operation check for vibration, shaft seal leakage, pump priming or need for venting, and heat build-up in the shaft seal, bearings, or pump and motor casings. The discharge pressure should be appropriate for the conditions.

503.4 Check to determine if the pump is delivering a flow which is consistent with that expected.

503.5 Use a stroboscope, a direct-reading tachometer, or other device selected during the test planning to determine if the design speed is attained.

503.6 When pressures, flow rates, speed, temperatures, and power readings are stable or within acceptable limits of fluctuation, proceed with the test run or runs necessary for the desired determinations.

504.0 *Test Run*

The data enumerated in this section will establish all the operating characteristics, but the data to be collected during a run may be abbreviated as decided in the test plan. Obtaining this bench-mark information on a new pump or a newly overhauled pump will provide the basis for evaluation of any future tests: any deficiencies observed in the trial run must be corrected. If none are observed, the test run can proceed.

504.1 Data required are suction pressure, discharge pressure, absolute pressure on the surface of-the liquid being pumped, liquid temperature at the suction, flow rate, power readings, and pump speed.

504.2 Line up and start the test systems as stated under the Trial Run Checks in Section 503.0, or proceed if the system is already, operating.

504.3 Establish the flow desired and allow conditions to stabilize.

504.4 Take the required readings and record on a data sheet, such as the example in the appendix (see Figure 802.2.1). The need for simultaneous data depends on whether readings are fluctuating. Obviously, if there is no change, one person can proceed from data point to data point and collect the readings.

504.5 Change the flow rate, then return to the first data point and allow time for the conditions to stabilize.

504.6 If taking data to establish characteristic curves the data points for increasing flow rates should be repeated with decreasing flows.

504.7 Check the pump, instrumentation and piping system after the test is completed to see that no changes have occurred which would cause erroneous readings.

600.0 Computation of Results

601.0 *Results Required*

The usual results desired are the computed coordinates based on the test data, which provide a comparison with the characteristic curves, issued by the pump manufacturer. (See Figure 702.2.1 characteristic curve.)

601.1 The coordinates needed for a point on the characteristic curves are total head and capacity, power and capacity, available net positive suction head and capacity, or efficiency and capacity.

602.0 *Total Head*

Total Head is the hydraulic energy added to the pumped fluid, i.e., the algebraic difference between the total discharge head and total suction head.

602.1 Readings of pressure gauges in customary (U.S.) units are converted to pounds per square inch absolute and then to feet of liquid head to obtain static head. Readings of SI Units are converted to kPa and then to meters of head of liquid. (See example in Section 803.0) Static head is defined in Section 200.0. Due regard must be paid to correcting for gauge location with regard to the chosen datum line.

602.2 The velocity used for determining the velocity head (suction or discharge), as defined in Section 200.0, may be obtained from the volume rate of flow and the cross-sectional area at the measurement section. Velocity is determined by:

$$v = \frac{Q}{448.8\,A}\ \text{for customary units}$$

$$v = \frac{Q}{A}\ \text{for SI units}$$

Note: 1 ft^3/ sec = 448.8 gpm

602.3 Flow to be used for determining liquid velocity may also be calculated from the following orifice flow equation:

$$Q = 448.8\, C\, A\, \sqrt{2gh} \text{ for customary units}$$
$$Q = C\, A\, \sqrt{2gh} \text{ for SI units}$$

C = meter coefficient (see reference 805.2 for typical coefficients)
h = differential head of fluid pumped as computed from the manometer head across the orifice
g = acceleration of gravity, 32.2 ft/ sec^2 or 9.81 m/sec^2 for SI units

603.0 *Capacity*

Capacity or volume rate of flow is in terms of cubic feet per second for calculation purposes and as gallons per minute for reporting purposes. The following equations may be used as they apply:

$$Q,\ \text{ft}^3/\text{sec} = \frac{\text{weight rate of flow, } \frac{lb}{sec}}{\text{specific weight, } \frac{lb}{ft^3}}$$

$$Q,\ \text{ft}^3/\text{sec} = \frac{\text{volume rate of flow, gpm}}{(60\,\text{sec/min})\ (7.48\ gal/ft^3)}$$

$$Q,\ \text{ft}^3/\text{sec} = (\text{velocity, ft/sec})\ (\text{area, ft}^2)$$

604.0 *Power*

The characteristic power curve is an indication of the rate of doing work applied to the pump.

604.1 Operating Driver Brake Horsepower or

$$H_p = \frac{(watts\ input)\ (motor\ efficiency)}{746}$$

32

Where a wattmeter is not used to obtain the input, the following is used on 3-phase motors:

watts input = (1.73) (phase voltage) (average phase amperage) (power factor)

604.2 Where other machines provide the power input, the manufacturer can supply means of obtaining this power. For example, a steam turbine driver might use nozzle block steam pressure, exhaust steam pressure and efficiency to determine the brake horsepower.

605.0 *Efficiency*

The pump power output combined with the efficiency (P_w/H_p) at any point may provide a check on the driver brake horsepower, or combined with the driver brake horsepower, used to determine efficiency.

605.1 Pump hydraulic power output is found by the following formula:

$$P_w = \frac{\text{(lb liquid pumped/min)(total developed head)*}}{33,000} \text{ for customary units}$$

$$P_w = \frac{\text{(kg liquid pumped/sec)(total developed head)*}}{101.9} \text{ for SI units}$$

* Not discharge pressure

If the liquid is of specific gravity (s) with reference to water at 69° F,

$$P_w = \frac{s\,Q\,H}{3,960} \text{ for customary units}$$

$$P_w = 9.81\ s\,Q\,H \quad \text{for SI units}$$

605.2 Pump efficiency is then determined as follows:

$$eff = \frac{output}{input} = \frac{P_w}{H_p}$$

606.0 *Speed*

Where the test speed, n_t is different than the speed used to develop the characteristic curves, the values of the coordinates may be estimated by the following equations:

$$Q = Q_t \left(\frac{n}{n_t} \right)$$

$$H = H_t \left(\frac{n}{n_t} \right)^2$$

$$\text{NPSH approximates* NPSH}_t \left(\frac{n}{n_t} \right)^2$$

$$P_w = P_{w\,t} \left(\frac{n}{n_t} \right)^3$$

where H = pump total head
n = speed of rotation

* for a specific pump design the exponent of the speed ratio may vary from 1.7 to 2.0 and can only be determined by testing.

Where the subscript "t" refers to the conditions at the test speed. Variations between the test speed and the characteristic curve speed that exceed 10% could affect the validity of the comparison.

607.0 *Impeller Diameter Variation*

Where the impeller diameter is different than the diameter used by the manufacturer to develop the characteristic curves, the values of the coordinates may be estimated by the following equations:

$$Q = Q_t \left(\frac{D}{D_t} \right)$$

$$H = H_t \left(\frac{D}{D_t} \right)^2$$

$$P_w = P_{w\,t} \left(\frac{D}{D_t} \right)^3$$

608.0 *Sample Test-Run Data Sheet*

Pump Type_____

Pump Size_____

Pump Manufacturer_____

Pump No. of Stages_____

Driver Type_____

Driver Speed_____rpm

Suction Temperature_____°F

Liquid Type_____

Specific Gravity_____

Barometric Pressure_____mm.Hg

Discharge Pressure♠_____psig

Motor Input Amps_____amps

Motor Power Factor_____

Turbine Nozzle Ring Pressure_____psig

Turbine Back Pressure_____psig

Pump Serial No._____

Rated Capacity_____

Rated Head_____

Impeller Diameter_____

Rated Horsepower_____

Driver Serial No. _____

Suction Pressure♠_____

NPSHR_____

Viscosity Absolute_____

Pressure (liquid surface) suction ____

Pressure (liquid surface) discharge__

Motor Input Volts_____

Motor Power Input_____

No. of Nozzles _____

♠ Corrected for elevation above or below datum elevation

System Sketch: (User to include the sketch for the actual application here)

700.0 Interpretation of Results

701.0 *Introduction*

The graphical presentation of data provides the most effective method of explaining results. Each curve should be identified and dated with the pertinent information such as pump serial number, impeller designation, impeller diameter, speed, and liquid pumped. Field test results are to be compared to the manufacturer's data. An understanding of the definitions of the following terms is essential.

1. Certified test curve
2. Certified curve
3. Published generalized curve

701.1 A certified test curve is a graphical representation of a physical test of a specific pump. Such tests were conducted under specific conditions and acceptance criteria, and were specified and paid for by the purchaser. The details of how the tests were be conducted, the acceptance criteria and the documentation of the results were defined to the purchaser prior to conducting the test. It should be noted that typically only the design point head, flow and efficiency were guaranteed and only within a specific tolerance. All other data is for information only. NPSH required was NOT guaranteed unless made a part of the purchase order and an NPSH required test was not conducted.

701.2 A certified curve is typically a copy of the published generalized curve (see definition below for this term) with the "design point" head and flow conditions noted on the curve. This document is a graphical representation of the contract requirements of the equipment which was supplied. The indicated conditions will be used as the acceptance criteria if the purchaser specified hydraulic tests were to be conducted. These curves are no more than the published generalized curves. During testing, alterations of the impeller diameter and vane entry surfaces may have been required to achieve the design point, contract conditions and fall within the specified acceptance tolerances.

701.3 Published generalized curves graphically represent a composite of a number of pump tests of the pump model indicated. There typically will be variation of the hydraulics actually produced by any specific impeller model and diameter from that which is illustrated on the curve. While the total head developed by the pump will normally be within an 8% tolerance of the specified values (with correspondingly different horsepower requirements), there can be substantially higher variations on certain pumps.

702.0 *Discussion*

702.1 Data obtained under field conditions may not represent a smooth curve. The curve represented by the data will be an average. If the data points deviate significantly from the average curve, investigation of the reasons for these deviations should be made.

702.2 There is a given relationship between horsepower, total head, capacity, and $NPSH_R$ requirements. As a crosscheck on the field test results, one parameter can be compared to the others using the manufacturer's published characteristic curves. If the head and horsepower are known, the capacity can be estimated. If the $NPSH_R$ is known, the capacity at that point can be estimated. If the horsepower is known, the capacity can be estimated. See the characteristic curve (Figure 702.2. 1) for the relationship between these parameters.

702.2.1 The relationship between the Power, total developed head and capacity for a given pump speed and impeller diameter may be expressed more rigorously by:

$$P_w = \frac{s\,Q\,H}{3{,}960\ (eff)} \text{ for customary units}$$

$$P_w = \frac{9.81\,s\,Q\,H}{(eff)} \text{ for SI units}$$

Where *eff* is the pump efficiency, defined in 605.2

Figure 702.2.1 illustrates the typical graphical performance data presentation of a centrifugal pump. Total head vs. capacity are illustrated for various impeller diameters operating at a fixed speed. $NPSH_R$ vs. capacity is also illustrated.

702.2.2 As speed or impeller diameter change, the total head (H), flow and power will change.

702.2.3   The actual measured test results at different speeds or impeller diameter other than illustrated by the manufacturer's data are to be adjusted per paragraph 606.0 and 607.0 of this document prior to comparing it to the manufacturer's published information.

702.3 Capacity measurements are usually the most difficult to obtain. In some field installations you may not find instrumentation for measurement of capacity. Indirect determination of capacity may be necessary. Changes in a known tank volume being supplied by the pump over a measured period of time may be the only field flow measurement available.

702.4 Head measurement is normally the easiest measurement to accomplish in the field. The differential pressure between the suction and the discharge at positions other than the inlet and outlet flanges of the pump measured with the instruments located at an elevation equal to the pump centerline will be required to be adjusted for elevation, static head, pressure drop due to friction losses and changes in velocity head.

702.5 The accuracy of field performance tests is directly dependent on the accuracy of the head and flow measurements. Pumps with a relatively flat performance curve (A), as illustrated by the solid line curve in figure 702.5.1 require more attention to the precision of the flow measurement to produce a meaningful performance curve. Curves with a steeper slope (B) as illustrated by the dotted line in figure 702.5.1 require attention to a precise head measurement.

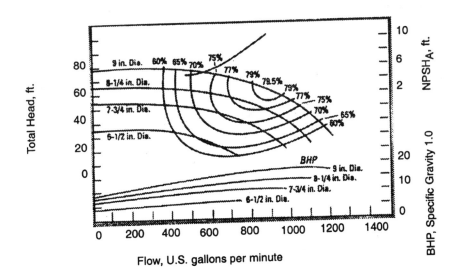

Figure 702.2.1 Typical Pump Performance Curve

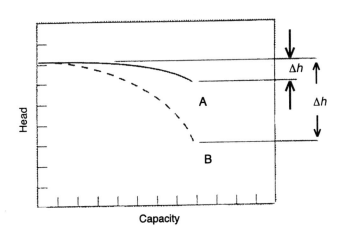

Figure 702.5.1 Capacity Vs. Head Curve

702.6 Comparing the field horsepower vs. flow to the corresponding data from the supplier will help evaluate major discrepancies that may exist in the pump's performance. Measured results will have to be adjusted to compensate for site condition, specific gravity and viscosity before making this comparison. The flat curve (A) in Figure 702.7.1 represented by the solid line would not permit such determination. Other methods would have to be used. The curve with dotted line (B) represents a very steep horsepower curve and would lend itself to indirect flow determination.

702.7 It should be noted that viscosity will affect the performance of a centrifugal pump. The curve in Figure 702.8.1 taken from the Hydraulic Institute Standards shows the effect of viscosity on the performance of a centrifugal pump. Viscosity characteristics of the fluid that is being pumped should be determined.

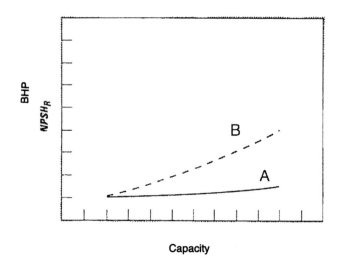

Figure 702.7.1 Capacity Vs. Horsepower Curve

702.8 Specific gravity will also have an effect on performance as shown in Figure 702.8.1. The specific gravity of the fluid must be known.

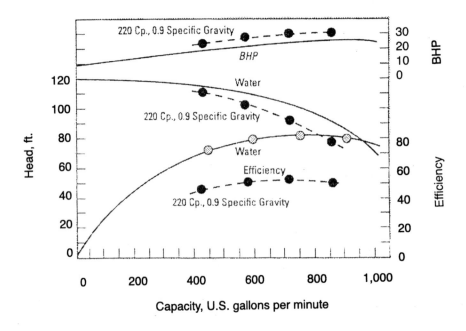

Figure 702.8.1 Typical Pump Curve

702.9 Changes in speed from those specified will affect the pump performance, as illustrated in Figure 702.9.1. Verification of speed is essential at each test point. Manufacturer's published curves illustrating the measured performance test results are adjusted to reflect constant driver speed throughout the range of flows. This speed will be noted on the performance curve. The measured field test results should be adjusted to this common speed using the information in paragraphs 606.0 and 607.0 of this procedure.

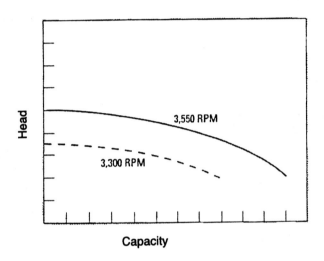

Figure 702.9.1 Capacity Vs. Head Curve with Speed as a parameter

703.0 *Trouble Checklist*

The following tables can be used to identify possible problem areas that can affect performance in centrifugal pumps.

Table 703.0.1 No Liquid Delivered

1. Priming - casing and suction pipe not completely filled with liquid.
2. Speed too low.
3. System head too high. Check the total head (particularly friction loss) or develop system curve.
4. Suction lift too high (suction pipe may be too small or long, causing excessive friction loss). Check with vacuum or compound gauge.
5. mpeller or suction or discharge pipe or opening completely plugged (i.e., valve shut off).
6. Wrong direction of rotation.
7. Air pocket in suction line.
8. Stuffing box packing worn or liquid seal plugged, allowing leakage of air into pump casing.
9. Air leak in suction line.
10. Not enough suction head for hot or volatile liquids. Check carefully as this is a frequent cause of trouble in such service.
11. Impeller key missing.
12. Broken shaft.
13. Pumps not properly vented.

Table 703.0.2 Not Enough Liquid Delivered

1. Priming - casing and suction pipe not completely filled with liquid.
2. Speed too low.
3. System head higher than anticipated. Check total head (particularly friction loss) or develop system curve.
4. Suction lift too high (suction pipe may be too small or long, causing excessive friction loss). Check with vacuum or compound gauge.
5. Impeller suction or discharge opening partially plugged.
6. Wrong direction rotation.
7. Air pocket in suction line.
8. Stuffing box packing worn or liquid seal plugged, allowing leakage of air into pump casing.
9. Air leak in suction line.
10. Not enough suction head for hot or volatile liquids. Check carefully as this is a frequent cause of trouble in such service.

11. Foot valve too small.
12. Foot valve or suction pipe not immersed deeply enough.
13. Mechanical defects:
    a. Impeller clearance too great.
    b. Impeller damaged.
    c. Wrong impeller (i.e., size).
14. Vortexing.
15. Entrained gases in liquid.
16. Impeller key missing.
17. Worn or missing wear rings.
18. Casing wear.

Table 703.0.3 Not Enough Pressure

1. Speed too low.
2. Air or gases in liquid.
3. Impeller diameter too small.
4. Mechanical defects:
    a. Impeller clearance too great.
    b. Impeller damaged.
5. Wrong direction of rotation.
6. Incorrect location of pressure gauge on discharge nozzle or discharge pipe.

Table 703.0.4 Pump Works a While and Then Quits

1. Leaky suction line.
2. Stuffing box packing worn-or liquid seal plugged allowing leakage of air into pump casing
3. Mechanical defects:
    a. Broken shaft.
    b. Broken coupling.
4. Motor stops due to overload trip out.
5. Vortexing or insufficient $NPSH_A$

Table 703.0.5 Noise

1. Insufficient $NPSH_A$
2. Bearings.
3. Foreign object in pump casings.
4. Vortexing.
5. Cavitation.

703.1 *Effects of Cavitation on performance testing:* Centrifugal pumps experience at least two types of cavitation. The first, and most pertinent to this procedure, is cavitation resulting from inadequate margin between the $NPSH_A$ and $NPSH_R$. This will cause a reduction in the pump hydraulic performance as well as, in time, mechanical damage to the pump. The second is cavitation resulting from what is known as "low flow recirculation".

703.2 Cavitation resulting from an inadequate NPSH margin or from too low a flow rate for the pump design must be eliminated before attempting to conduct field-tests which will be compared to the manufacture's data. NOTE: Severe and extended periods of cavitation can damage the pump and should be eliminated. See reference 805.1.

703.2.1 The presence of moderate to severe cavitation can normally be determined by the existence of a crackling sound emanating from the suction side piping of the pump. It is advisable that a good quality mechanic's stethoscope be used to conduct this evaluation.

703.2.2 At flow rates near or below the manufacturer's published or tested NPSH values, the noise of recirculation cavitation may appear, but will not affect performance test results. If cavitation noise disappears at higher flow rates, the cavitation can be attributed to the recirculation phenomena and the test effort continued without any change to the system.

703.2.3 The presence of cavitation may also be determined by a strong axial vibration component that disappears with a larger NPSH margin shown in Figure 703.2.3. A broad band vibration frequency will also typically exist when performing a vibration spectrum analysis of the pump. The greater the cavitation the larger the amplitude of the vibration will exist over the broad band range.

703.3 If the purpose of a field test is to compare the pump performance to the "as new condition" and to the supplier's published or test results, it will be meaningless to proceed without first eliminating the cavitation due to an inadequate NPSH margin.

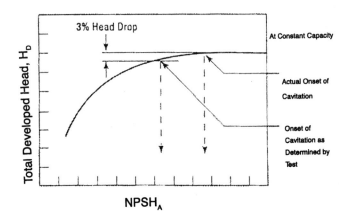

Figure 703.2.3 NPSH on the Pump Curve

703.4 Reduction and Control of Cavitation Damage If a given pump has suffered damage from cavitation and it is not possible to increase the $NPSH_A$ or reduce $NPSH_R$ by means of operating at a lower flow, certain things can be done to alleviate or reduce continued damage short of massive modifications such as lowering the pump.

703.4.1 Certain materials have relatively great resistance to cavitation attack. Cast iron and other soft materials are especially prone to damage. It may be possible to replace pump impeller and/ or casing with a material which has a much higher resistance to damage.

703.4.2 After consulting with the manufacturer the leading edge of the impeller vanes can be cut back and perhaps the impeller eye enlarged thereby reducing velocity and its attendant pressure drop. The pump manufacturer should be contacted concerning an alternative impeller pattern that includes these modifications that may still perform with satisfactory head, capacity, and efficiency.

703.4.3 The existing impeller vanes may be sharpened by filing to a narrow leading edge by the manufacturer.

703.4.4 Pre-rotation in the suction piping can be reduced by installation of a simple straightening vane. Care should be taken to insert vanes with a minimum pressure drop.

703.4.5 Impeller inlet vanes and casing can be coated with a resilient, nonmetallic coating material that provides a cushioning effect to the shock waves produced by the collapsing bubbles. Applying such coatings is a special procedure, will probably affect the head/capacity performance of the pump and may not be possible on high-speed units.

703.4.6 For materials near their boiling point the suction line may be cooled to lower the vapor pressure, thereby raising the $NPSH_A$.

800.0 Appendix
801.0 *Nomenclature*

Term	Symbol	Customary Units	S.I. Units
Area	A	ft^2	m^2
Impeller Diameter	D	inches	m
Voltage	E	volts	V
Pump Efficiency	Eff	%	%
Local Gravitational Acceleration	g	ft/sec^2	m/sec^2
Total Head	H	feet of liquid	m of liquid
Total Discharge Head	h_d	feet of liquid	m of liquid
Discharge Friction Head	h_{fd}	feet of liquid	m of liquid
Suction Friction Head	h_{fs}	feet of liquid	m of liquid
Total Suction Lift	h_{gs}	feet of liquid	m of liquid
Power	Hp	horsepower	kW
Pressure Head	h_p	feet of liquid	m of liquid
Suction Surface Pressure	h_{ps}	feet of liquid	m of liquid
Suction Surface Pressure Absolute	h_{psa}	psia	kPa
Total Suction Head	h_s	feet of liquid	m of liquid
Static Discharge Head	h_{sd}	feet of liquid	m of liquid
Static Suction Head	h_{ss}	feet of liquid	m of liquid
Velocity Head	h_v	ft of liquid	m of liquid
Discharge Velocity Head	h_{vd}	feet of liquid	m of liquid
Vapor Pressure	h_{vpa}	psia	kPa
Suction Velocity Head	h_{vs}	feet of liquid	m of liquid
Current	I	amps	A
Length	L	ft	m
Speed of Rotation	n	rpm	rad/sec
Net Positive Suction Head Available	NPSH$_A$	ft of liquid	m of liquid
Net Positive Suction Head Required	NPSH$_R$	ft of liquid	m of liquid
Suction Pressure	P_1	psig	kPa
Discharge Pressure	P_2	psig	kPa
Power Factor	Pf	dimensionless	dimensionless
Gauge Pressure	P_g	psig	kPg
Hydraulic Power	Pw	horsepower	kW
Capacity of Pump	Q	gpm	m^3/sec
Specific Gravity	s	dimensionless	dimensionless
Temperature	T	°F	°C
Velocity	v	ft/sec	m/sec
Work	W	ft lb/ lb	Joules/ Kg
Datum Correction Discharge	Z_d	feet of liquid	m of liquid
Datum Correction Suction	Z_s	feet of liquid	m of liquid
Specific Weight (Mass)	γ	lb/ft^3	kg/m3
Viscosity	μ	centipoise	Pa

802.0 *Sample Test Results*

The example problem is shown to demonstrate calculations needed to establish various pump characteristics at a single operating point. When complete performance curves are required, several points must be determined.

802.1 *System Configuration* See Figure 802.1

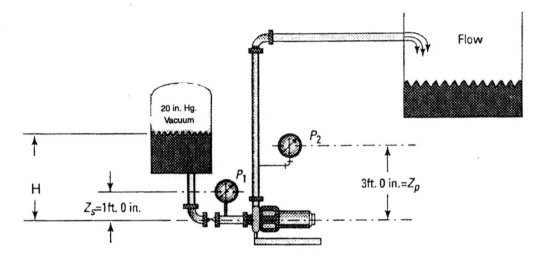

Figure 802.1 System Schematic

802.2 *Sample Data Log Sheet* Figure 802.2.1 is an example of a log sheet used to assemble data, which will be used in the sample calculation. Sample calculation is based on only one flow condition (1000 gpm). All other flow conditions will be taken from Figure 802.2.2.

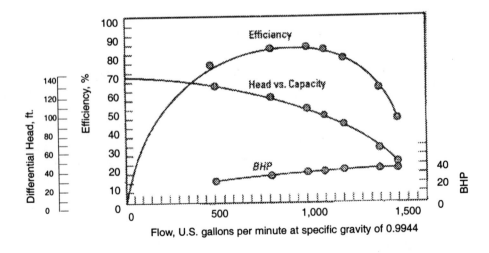

Figure 802.2.1 Vendor Supplied Pump Curve

802.3 Flow Measurement was determined by measuring the difference in level in the discharge tank over a time period.

802.4 *Power for Sample Problem* The power factor shown in the data was obtained from the motor manufacturer's typical values for the type and size motor. The motor efficiency was obtained in the same manner.

RECORD OF PUMP TEST

Mark each pump with test number

MANUFACTURER	DISCHARGE DIAMETER	SUCTION DIAMETER	IMPELLER DIAMETER	TEST NO.
BRAND X	3"	4"	12"	1

MODEL DESCRIPTION	LIQUID TEMP.	BAROMETRIC PRESSURE	SPECIFIC GRAVITY	DATE
	100° F	14.7	0.9944	11-17-2001

MANUFACTURER SERIAL NO.	G.P.M. AT	FT. TOTAL HEAD	VAPOR PRESSURE	VISCOSITY	TESTED BY
	1000	98	0.947	1 cp	

MOTOR (DRIVER)	MOTOR SPEED		LINE VOLTAGE
40.0 HP MOTOR	1750 RPM		460 VOLTS

EFFICIENCY %	POWER FACTOR	CYCLES/HERTZ	PHASE
90	0.875	60	3

RATE OF FLOW GPM	P_1 SUCTION PRESSURE GA. PSIA	P_2 DISCHARGE PRESSURE GA. PSIA	Z_d DISCHARGE GA. CORRECTION FEET	Z_s SUCTION GA. CORRECTION FEET	TOTAL HEAD FEET	POWER INPUT AMPS	BRAKE H.P. HP	PUMP SPEED RPM	CORRECTED FLOW GPH	CORRECTED HEAD FEET	CORRECTED B.H.P. HP	WATER H.P. WHP	PUMP EFF. %
0	11	68	3.0	1.0	135	18	15	1750	None	None	None	0	0
500	10.5	63			125	26.2	21.6	1745				15.8	73
800	10	57			112	31	27.6	1749				22.6	82
1000*	9.0	49			96	36	30	1750				24.9	83
1100	8.5	46			90	37	30.8	1747				25	81
1200	8	42			80	37.3	31.1	1752				24.3	78
1400	7	30			55	39	32.4	1750				19.4	60
1500	6.3	22			40	41.5	34.4	1753				15.1	44

REMARKS: Piping:
Suction — 6" SCH 40 STEEL
Discarge — 6" SCH 40 STEEL

* Condition used in sample calculation, Sec. 803.0

Figure 802.2.2 Recording Sheet for Pump Test

803.0 *Sample Calculations (Dual Units)*

<table>
<tr><td colspan="2" align="center"><u>U.S. Customary Units</u>
Pressure Head</td></tr>
</table>

U.S. Customary Units — Pressure Head

$$h_p = \frac{P_g \frac{lb_f}{ft^2} \; 32.2 \frac{lb_m \, ft}{lb_f \, sec^2}}{\gamma \frac{lb_m}{ft^3} \; g \frac{ft}{sec^2}}$$

$(144) \, P_g \, (PSIA) = P_g \, (lb_f / ft^2)$

$(62.4) \, s = \gamma \, (lb_m / ft^3)$

$g = 32.2 \, (ft / sec^2)$

$$h_p = 2.31 \frac{P_g (PSIA)}{s} ft$$

Flow Rate
$Q = 1000 \, gal / min$

Static Suction Head, h_{ss}
$P_1 = 9 \, PSIA$

$$h_{ss} = 2.31 \frac{P_1(PSIA)}{s} ft$$

$s = 0.9944$

$h_{ss} = 2.31 \, (9) \, / \, (0.9944)$
$= 20.91 \; feet$

Static Discharge Head, h_{sd}
$P_2 = 49.3 \, PSIA$

$$h_{sd} = 2.31 \frac{P_2(PSIA)}{s} ft$$

$h_{sd} = 2.31 \, (49.3) \, / \, (0.9944)$
$= 114.52 \; ft$

S.I. Units — Pressure Head

$$h_p = \frac{P_g \frac{N}{m^2} \; 1 \frac{kg \, m}{N \, sec^2}}{\gamma \frac{kg}{m^3} \; g \frac{m}{sec^2}}$$

$(1000) \, P_g \, (kPa) = P_g \, (N / m^2)$

$(1000) \, s = \gamma \, (kg / m^3)$

$g = 9.807 \, (m / sec^2)$

$$h_p = 0.102 \frac{P_g (kPa)}{s} \; meters$$

Flow Rate

$$Q = 1000 \frac{gal}{min} \; \frac{0.003785 \, m^3}{gal} \; \frac{60 \, min}{hr}$$

$= 227.1 \, m^3 / hr$

Static Suction Head, h_{ss}

$P_1 = 9 \, PSIA \frac{6.894 \, kPa}{PSIA}$

$= 62.05 \, kPa$

$$h_{ss} = 0.102 \frac{P_1(kPa)}{s} \; meters$$

$s = 0.9944$

$h_{ss} = 0.102 \, (62.05) \, / \, (0.9944)$
$= 6.36 \; meters$

Static Discharge Head, h_{sd}

$P_2 = 49.3 \, PSIA \frac{6.894 \, kPa}{PSIA}$

$= 339.87 \, kPA$

$$h_{sd} = 0.102 \frac{P_2(kPa)}{s} \; meters$$

$h_{sd} = 0.102 \, (339.87) \, / \, (0.9944)$
$= 34.86 \; meters$

U.S. Customary Units	**S.I. Units**

Suction Velocity Head, h_{vs}

$h_{vs} = \dfrac{V^2}{2g}$ for 6" line

$V = 11.1$ ft / sec

$h_{vs} = \dfrac{(11.1 \text{ft / sec})^2}{2\,(32.17 \text{ ft / sec}^2)}$

$= 1.91$ ft

Discharge Velocity Head, h_{vs}

$h_{vd} = \dfrac{V^2}{2g}$ for 6" line

$V = 11.1$ ft / sec

$h_{vd} = \dfrac{(11.1 \text{ ft / sec})^2}{2\,(32.17 \text{ ft / sec}^2)}$

$= 1.91$ ft

Datum Correction Suction, Z_s

$Z_s = 1$ ft

Datum Correction Discharge, Z_d

$Z_d = 3$ ft

Vapor Pressure, h_{vpa}

$h_{vpa} = 0.947$ PSIA

$h_{vp} = 2.31 \dfrac{P_{vpa}(\text{PSIA})}{s} \text{ft}$

$h_{vp} = 2.31\,(0.947) / (0.9944)$
$= 2.199$ ft

Total Suction Head, h_s

$h_s = h_{ss} + Z_s + h_{vs}$
$= 20.91 + 1.0 + 1.91$
$= 23.82$ ft

Suction Velocity Head, h_{vs}

$h_{vs} = \dfrac{V^2}{2g}$

$V = 11.1$ (ft / sec) (0.3048 m / ft)
$= 3.38$ m / sec

$h_{vs} = \dfrac{(3.38 \text{m / sec})^2}{2\,(9.807 \text{ m / sec}^2)}$

$= 0.584$ m

Discharge Velocity Head, h_{vs}

$h_{vd} = \dfrac{V^2}{2g}$

$V = 11.1$ (ft / sec) (0.3048 m / ft)
$= 3.38$ m / sec

$h_{vd} = \dfrac{(3.38 \text{m / sec})^2}{2\,(9.807 \text{ m / sec}^2)}$

$= 0.584$ m

Datum Correction Suction, Z_s

$Z_s = 1$ ft (0.3048 m / ft) $= 0.3048$ m

Datum Correction Discharge, Z_d

$Z_d = 3$ ft (0.3048 m / ft) $= 0.91$ m

Vapor Pressure, h_{vpa}

$h_{vpa} = 0.947$ PSIA (6.894 kPa/PSIA)
$= 6.529$ kPa

$h_{vp} = 0.102 \dfrac{P_{vpa}(\text{kPa})}{s} \text{meters}$

$h_{vp} = 0.102\,(6.529) / (0.9944)$
$= 0.670$ m

Total Suction Head, h_s

$h_s = h_{ss} + Z_s + h_{vs}$
$= 6.36 + 0.3048 + 0.548$
$= 7.248$ m

U.S. Customary Units | ## S.I. Units

Total Disharge Head, h_d

$$h_d = h_{sd} + Z_d + h_{vd}$$
$$= 114.52 + 3 + 1.91$$
$$= 119.43 \text{ ft}$$

Total Disharge Head, h_s

$$h_d = h_{sd} + Z_d + h_{vd}$$
$$= 34.86 + 0.91 + 0.584$$
$$= 36.36 \text{ m}$$

Total Head, H

$$H = h_d - h_s$$
$$= 119.43 - 23.82$$
$$= 95.61 \text{ ft}$$

Total Head, H

$$H = h_d - h_s$$
$$= 36.36 - 7.248$$
$$= 29.11 \text{ m}$$

Driven Horsepower

$$Hp = \frac{\sqrt{3} \text{ E I Pf Eff}}{746 \text{ W hp}}$$

$$= \frac{\sqrt{3}(460)(36)(0.875)(0.90)}{746 \text{ W hp}}$$

$$= 30.28 \text{ hp}$$

Driven Horsepower

$$Hp = \sqrt{3} \text{ E I Pf Eff}$$

$$= \sqrt{3}(460)(36)(0.875)(0.90)$$

$$= 22588 \text{ W} = 22.59 \text{ kW}$$

Hydraulic Horsepower

$$P_w = Q H \gamma$$

$$Q = 1000 \text{ gal / min (ft}^3 / 7.48 \text{ gal)}$$

$$= 133.7 \text{ ft}^3 / \text{min}$$

$$= 2.228 \text{ ft}^3 / \text{sec}$$

$$H = 95.51 \text{ ft}$$

$$\gamma = 0.9944 (62.4 \text{ lb / ft}^3)$$

$$= 62.05 \text{ lb / ft}^3$$

$$P_w = (2.228)(95.51)(62.05)$$

$$= \frac{13204 \text{ ft lb / sec}}{550 \text{ (ft lb / sec) / hp}}$$

$$= 24.01 \text{ hp}$$

Hydraulic Power

$$P_w = Q H \gamma g$$

$$Q = 227.1 \text{ m}^3 / \text{hr (hr / 3600 sec)}$$

$$= 0.06308 \text{ m}^3 / \text{sec}$$

$$H = 29.10 \text{ m}$$

$$\gamma = 0.9944 (1000 \text{ kg / m}^3)$$

$$= 994.4 \text{ kg m}^3$$

$$g = 9.807 \text{ m / s}^2$$

$$P_w = (0.06308) (29.10) (994.4) (9.807)$$

$$= 17901 \text{ kg m}^2 \text{ s}^3$$

$$= 17.90 \text{ k W}$$

U.S. Customary Units	S.I. Units
Efficiency	**Efficiency**

$$Eff = \frac{P_w}{H_p}(100\%)$$

$$= \frac{24.01 \text{ hp}}{30.28 \text{ hp}}(100\%)$$

79.3%

$$Eff = \frac{P_w}{H_p}(100\%)$$

$$= \frac{17.90 \text{ kw}}{22.59 \text{ kw}}(100\%)$$

79.3%

NPSH$_A$

$$\text{NPSH}_A = h_{ss} + Z_s - h_{vp}$$

$$= \quad 20.91 + 1 - 2.199$$

$$= \quad 19.7 \text{ ft}$$

NPSH$_A$

$$\text{NPSH}_A = h_{ss} + Z_s - h_{vp}$$

$$= \quad 6.36 + 0.3048 - 0.670$$

$$= \quad 5.99 \text{ m}$$

804.0 *Related Calculations*

The following calculations can be used to determine conditions, which affect pump performance.

804.1 *Net Positive Suction Head Available,* $NPSH_A$ should be compared with Net Positive Suction Head Required. $NPSH_R$ from the pump manufacturer's curve or data. Insufficient head can cause pumping problems.

804.2 *Effects of Speed and Impeller Diameter* The effects of minor changes in speed or impeller diameter can be computed when test conditions vary from published data.

804.2.1 Constant Speed Capacity varies in direct proportion to the impeller diameter:

$$Q_2 = Q_1 \left(\frac{D_2}{D_1} \right)$$

Total Head varies as the square of the ratio of the impeller diameters:

$$H_2 = H_1 \left(\frac{D_2}{D_1} \right)^2$$

Horsepower varies as the cube of the impeller diameter ratio:

$$Hp_2 = Hp_1 \left(\frac{D_2}{D_1} \right)^3$$

804.2.2 Constant Impeller Diameter Capacity varies in direct proportion to the speed:

$$Q_2 = Q_1 \left(\frac{n_2}{n_1} \right)$$

Total Head varies as the square of the ratio of the speeds:

$$H_2 = H_1 \left(\frac{n_2}{n_1} \right)^2$$

Horsepower varies as the cube of the speed ratio:

$$Hp_2 = Hp_1 \left(\frac{n_2}{n_1} \right)^3$$

804.2.3 Where:

D = *Impeller Diameter*
H = *Head*
Q = *Capacity*
n = *Rotational Speed*
Hp = *Power*

805.0 *References*

805.1 *Hydraulic Institute Standards available from Hydraulic Institute, 9 Sylvan Way, Parsippany, NJ 07054-3802*

805.2 *Cameron Hydraulic Data, 18th Edition: Copyright 1996, Available from Ingersoll-Rand Co., Woodcliff Lake, NJ 07675.*

805.3 *American National Standards ASME Centrifugal Pumps Performance Test Code. PTC 8.2. available from American Society of Mechanical Engineers, United Engineering Center, 3 Park Avenue, New York, NY 10016-5901.*

805.4 *American National Standards ASME Centrifugal/ Vertical Pumps; Allowable Operating Region 9.6.3. Available from American Society of Mechanical Engineers, United Engineering Center, 3 Park Avenue, New York, NY 10016-5901.*